AF496458

Comprehension Lifters

Book Two

High-interest activities for students with Special Needs.

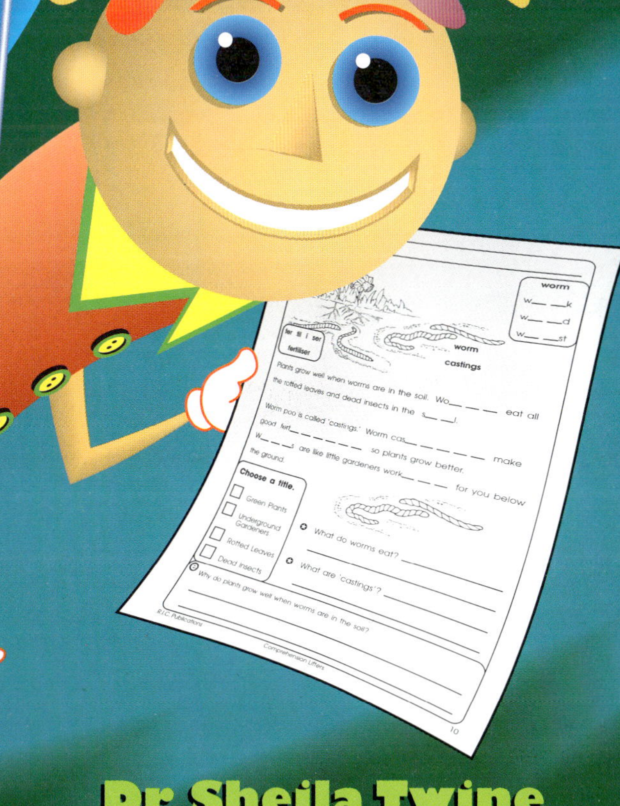

Dr Sheila Twine

PR–0372
ISBN 1-86400-305-7

Copyright Information

Comprehension Lifters – Book 2
Prim-Ed Publishing

First published in 1998 by R.I.C. Publications
Reprinted under license in 1999 by Prim-Ed Publishing

Copyright Dr Sheila Twine 1998

This master may only be reproduced by the original
purchaser for use with their class(es). The publisher prohibits
the loaning or onselling of this master for the purposes of
reproduction.

ISBN 1 86400 305 7
PR–0372

Additional titles available in this series are:
Comprehension Lifters – Book 1
Comprehension Lifters – Book 3
Comprehension Lifters – Book 4

Prim-Ed Publishing Pty. Ltd.
Offices in: United Kingdom: PO Box 051, Nuneaton, Warwickshire, CV11 6ZU
                    Australia: PO Box 332, Greenwood, Western Australia, 6024
                    Republic of Ireland: PO Box 8, New Ross, County Wexford, Ireland

Comprehension Lifters

Written by Dr Sheila Twine

Published by Prim-Ed Publishing

Foreword

Comprehension Lifters is a set of copymasters designed to assist students who have experienced difficulties with reading and comprehension. Topics are of high interest to appeal to older students with low literacy skills. Comprehension activities are varied and include superficial and in-depth questions requiring students to interact with text at a range of levels.

The four books of *Comprehension Lifters* can be used on their own or in conjunction with the four books of *Literacy Lifters* by the same author. Thus, students working at a basic level on Book 1 of *Literacy Lifters* would progress to Book 1 of *Comprehension Lifters* before moving on to Book 2 of *Literacy Lifters*.

In both *Literacy Lifters* and *Comprehension Lifters*, the book numbers do not correspond to grade or class levels, but rather to the skill level of students. Books progress in level of difficulty. Thus, Book 1 is pitched at a basic reading level and would be suitable for students who had floundered with their start to literacy. Book 2 would be appropriate for students who require assistance with basic skills. Books 3 and 4 would suit students whose reading and understanding of text is at a low level.

About the Author

Dr Sheila Twine is an educational consultant who has worked with parents, teachers and students with special needs in England, Scotland and Australia. She is the author of three books containing practical techniques for working with children who experience difficulties with reading and spelling. She holds a Masters Degree and a Doctorate in Education.

Dr Twine has been president of various associations and foundations involved with underachieving children with a variety of disabilities from mild intellectual handicap to attention deficit disorder. She was principal of a residential remedial primary school and has been the director of an education consultancy for many years.

Contents

GENERAL LAYOUT

Comprehension Lifters are designed as a teaching tool to assist you in raising the literacy levels of your students who are experiencing difficulties with reading and comprehension.

The page layout is similar throughout the books with text well separated to make discouraged readers feel more at ease. In each book there are 40 topic pages dealing with aspects of the same theme which are intended to be of interest to older students with low literacy skills. The font size is appropriate to the students' ages and so is smaller than normal for the easier reading level; therefore, the pages do not look 'babyish'.

The **Backing Sheet** can be copied onto the back of any or all of the topic pages. It is general and designed to complement your teaching. It contains space for activities in word study, thought-webbing and writing.

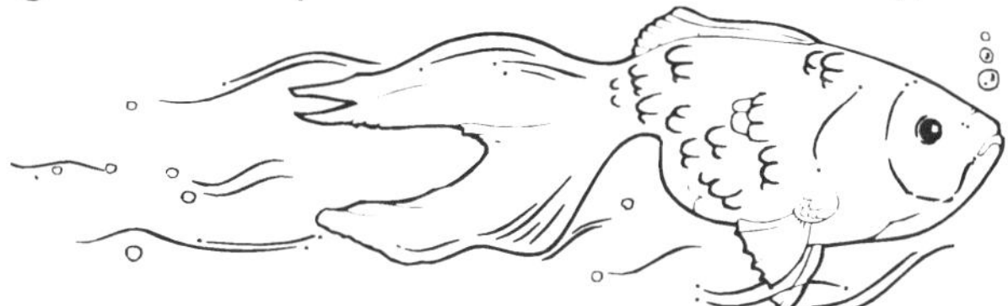

TOPIC PAGES

Each topic page has a number of sections. There is a section for patterned words which are phonically regular (found, sound, round) and for sight words (the, said, enough) for instant recognition. Most pages also have longer words broken into 'chunks' to encourage your students to tackle unfamiliar words more easily (nav-i-gate).

There are many comprehension sections. These include:

Main idea – Students need to be able to group the main idea of the text. Selecting and creating a title assists with this and is featured on every topic page.

Cloze activity – Each page also features a cloze activity which requires students to think as they read so a sensible word can be inserted into the blank spaces using clues from the text.

Questions – **Literal** questions are included where the answer is clear from the story. There are also **inferential** questions where students are asked to think and provide a logical answer. (He put on his waterproof coat – What was the weather like?)

Others – Some pages feature **following directions** or **giving opinions**. Some ask for students to **make judgements** and some ask for elements to be put in correct **sequence**.

At the bottom of each topic page there is a section in smaller type. This can act as a challenge to the more able students in your group to read to the others. The section either asks a deeper level of comprehension question or gives more information about the topic.

TEACHING TIPS

Two teaching strategies which were used when the layout of the topic pages was being trialled concerned **pre-reading** and **gap-filling**. You may like to try them.

Pre-reading – students were asked to glance at the pictures on their topic page then turn the page over (so no reading took place). They were then asked to volunteer snippets of information on the topic.

(It's all about ballooning, My Mum went up in a hot-air balloon, They have a fire machine to make the balloon go up, Some balloons go on races etc.)

This serves to get your students' brains tuned in to the topic. Three things then happen. First, the reading becomes easier (they're expecting the words). Second, their comprehension is better, and third, recall is improved.

Three good reasons for spending a few moments in a pre-reading activity.

Gap-filling – the topic pages are teaching tools for you so it's a good idea not to let your students fill in the blanks while you are going through the various sections of the page with them. The 'pencils down' rule allows them to focus their whole attention on what you're saying. Later comes their turn for practice and activity by filling in the blanks and answering the questions on their own. The **Backing Sheet** can be used while you're teaching. For instance,

'Let's spell that word together. Now spell it silently, turn over and see if you can write it down'.

Teacher Information

Topic Pages

The **Topic Pages** have been designed for students who are experiencing difficulties with literacy.
They are all pitched to create a high level of interest which will appeal to students with low skill levels.

The **Topic Page** layouts have been trialled in small remedial groups using an ACTIVE teaching mode which is outlined in the model below. Students were encouraged to fill in 'gaps' in the **Topic Pages** only after teaching had taken place. The **Backing Sheet** was used during the teaching for students to write patterned, chunked and sight words from memory. Topics are arranged in themes and each page contains scope for your active teaching as follows:

Pre-reading

- Discovering pre-existing knowledge through discussion, with students volunteering snippets of information after looking at illustrations.

Comprehension

- Predicting.
- Main idea – creating or choosing titles.
- Cloze activity – to promote thinking and to reinforce word study items.
- Questions – literal and inferential.

Word Study

- Phonic word patterns of regular words or theme words.
- Chunked words – words broken into chunks to assist blending and spelling rather than conventional syllables.
- Silent letters have been identified by slipping the letter below the line of the whole word.

Reading

- Oral, group, silent, paired or partner reading.

Minibeasts
Title:
worm
w___ ___k
w___ ___d
w___ ___st
worm
castings
fer til i ser
fertiliser
Plants grow well when worms are in the soil. Wo___ ___ ___ eat all
the rotted leaves and dead insects in the s___ ___l.
Worm poo is called 'castings.' Worm cas___ ___ ___ ___ ___ make
good fert___ ___ ___ ___ ___ ___ so plants grow better.
W___ ___ ___s are like little gardeners work___ ___ ___ for you below
the ground.
Choose a title.
Green Plants
Underground Gardeners
Rotted Leaves
Dead Insects
What do worms eat?
What are 'castings'?
Why do plants grow well when worms are in the soil?

Extension – Reading or Comprehension

Either a more difficult level of comprehension or more information on the topic.

Backing Sheet

The **Backing Sheet** is provided for you to copy on the back of any or all of the **Topic Pages**.
You'll notice that it is general and suitable for all **Topic Pages**. It is designed to complement your teaching
and you may care to use or adapt the following suggestions as shown on the sample **Backing Sheet**.

Patterned Words

- Can be used while you're teaching or for practice afterwards.

Odd Words/Long Words

- Practice in reading and writing of longer words or sight words.

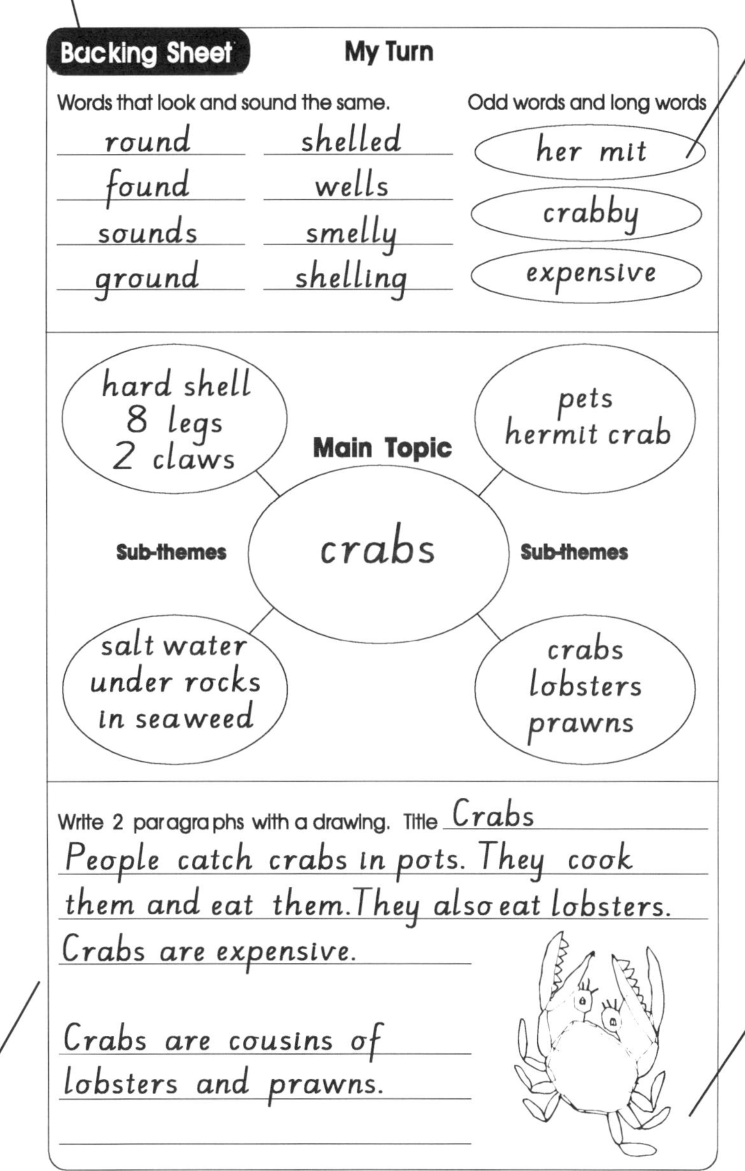

Drawing

- Matching drawing to main idea of story.

Writing

- Written (factual) material can assist students to write in a logical, thematic structure by giving them a standard format to formulate good habits.

- Later, format can be extended as basic skill improves and is well founded.

Thought Web

- Integrating elements from text with pre-existing knowledge.

- Practice in sorting out random thoughts into sub-themes, or lists or categories.

Words that look and sound the same.   Odd words and long words

______________   ______________

______________   ______________

______________   ______________

______________   ______________

Main Topic

Sub-themes   **Sub-themes**

Write 2 paragraphs with a drawing.   Title ______________

____

____

____

____

____

Page	Subject	Word Study	Feature	Word Count
1	Ladybirds	I'm, We're, It's	absurdities	37
2	Rats	I've, I'm, ole	picture distraction	41
3	Ants	theme words	draw from description	43
4	Mongooses	ey	write titles	34
5	Postcard	ay	inference, clues	45
6	Letter	air	using clues	43
7	Daddy-long-legs	first, second	picture distraction	43
8	Dragonflies	breath/breathe	true/false	42
9	Prawns	first, second, third	true/false	41
10	Worms	wor	sequence	46
11	Crabs	use, using	inference	38
12	Budgies	any, many	inference	43
13	Tadpoles	tch	draw from instructions	42
14	Starfish	all	yes/no	46
15	Fleas	ea	absurdities	48
16	Lizards	break, broken	inference	51
17	Baby Kangaroos	lk	true/false	52
18	Termites	ou	true/false	49
19	Beetles	synonyms	true/false	46
20	Goldfish	old	inference	40
21	Crickets	ass	confusions	50
22	Bees	ve	inference	50
23	Moles	ind, under	yes/no/doesn't say	47
24	Snails	nch	think up titles	50
25	Scorpions	ey	like/different	49
26	Tortoises	theme words	I'm a ......, I'm not a ...	58
27	Centipedes	ever, never	riddles	53
28	Guinea Pigs	o-e	think up a title	56
29	Poodles	oy	using clues	58
30	Hermit Crabs	bby, gger	inference	51
31	Frogs	ies	poem	59
32	Quails	qu, ey	picture inference	54
33	Bats	igh, some, come	what are we?	48
34	Mosquitoes	wh	poor title	55
35	Butterflies	tty, tter, tting	absurdities	67
36	Caterpillars	numbers	what am I?	53
37	Mice	gnaw, know	what am I?	72
38	Piglets	ood, other	what am I?	67
39	Cockroaches	could, would, should	what am I?	66
40	Baby Alligators	tch	what am I?	71

Title: ___________________________________

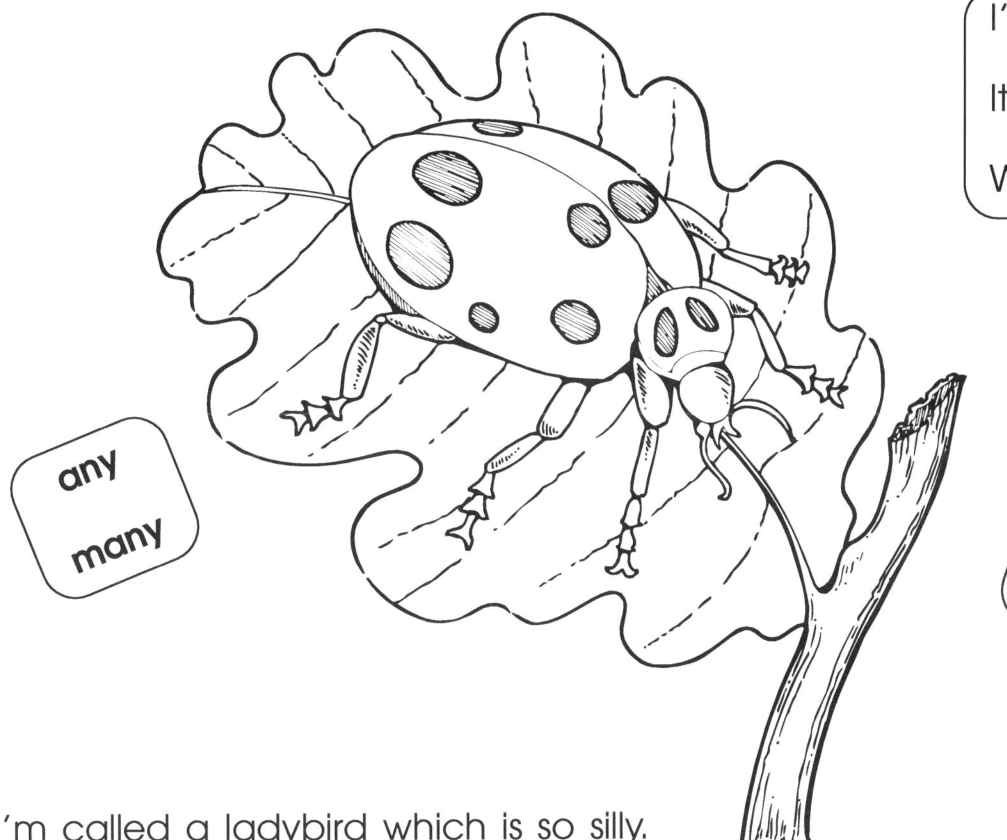

I'm called a ladybird which is so silly.

I'__ not a bi__d as anyone can see.

We'__ __ not all lad__ __s. Half of us are

females and h__ __f are males.

Where did we get such a si__ __y name?

Think of some better names for us.

____

____

____

Write a title at the top of the page.

⭐ A ladybird is a female bird. | Yes | No |

⭐ Half the ladybirds are male. | Yes | No |

e Think of a little story about how the ladybird got its name.
You could write it on the back of this page.

Title: ___________________________

_ _ _ _e

mol___
hol___
lik___
mak___
tak___

I'm
I am
I've
I h_ _ _ _

dif fer ent
different

I'm a rat. I'm a bit lik___ a mouse but I'm big___ ___ ___.

I'm a bit lik___ a mol___ but I've got a tail and I see well.

I'___ not at all like a hen but I l___ ___e to eat their f___ ___d.

Choose a title.

☐ Hens

☐ Seeds

☐ Mouse

☐ Rat

✪ How am I dif___ ___ ___ ___ ___t from

a mouse? ___________________________

✪ Do I like seed? ___________________________

✪ Does a mole see well? ___________________________

✪ What was your clue? ___________________________

____

ⓔ Is the picture a good one to explain the story? ___________________________

What would you have drawn? ___________________________

____

Title: _______________________________

either
neither

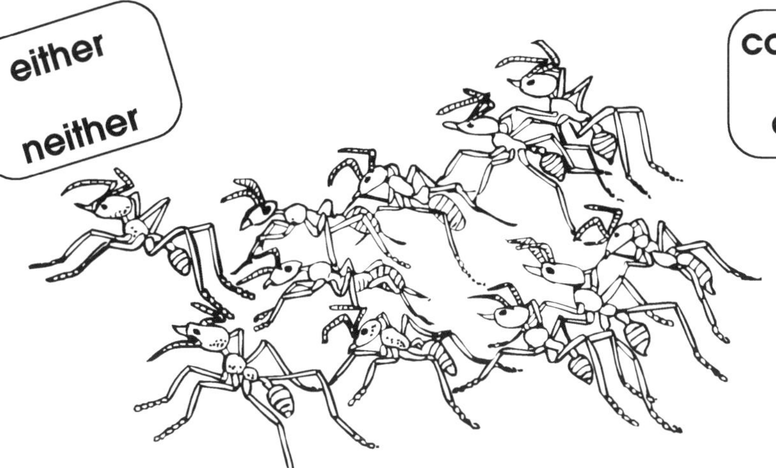

col on ies

colonies

birds

wasps

termites

tunnels

Ants live in col___ ___ies in underground homes called nests. No, not that sort of nest, that's for b___ ___ds.

No, not that s___ ___ ___ of nest eith___ ___. That's for wa___ ___s.

Ants live in a network of underg___ ___ ___ ___ ___ tun___ ___ ___ ___ and rooms. That is an ants n___ ___ ___.

Draw an ants nest.

⭐ Is this an ants nest or a termites nest? _______________

⭐ Think up a good title.

____

⭐ Who lives in a nest? _______________,

_______________, _______________, _______________

e

Think of a better name than 'nest' for an ants home.

____

____

Title: _______________________

mongoose

ey

th__ __

gr__ __

ob__ __

own er

The mon__ __ __ __ __ is a small brown

or gr__ __ animal. It can be very tame and

will ob__ __ its own__ __.

In India these little anim__ __ __ attack

and kill poi__ __ __ __ __ __ snakes.

Th__ __ are fast and fearless.

India

poi son ous

poisonous

Write two titles.

____

____

____

⭐ What colours are mongoose fur?

_______________ and _______________

⭐ What does 'fearless' mean?

____

____

e Why do you think it would be good to have a pet mongoose if you

lived in India? _______________________

____

Title: ___________________________

windy

ay

d__ __

pl__ __ing

spr__ __

Sund__ __

Mond__ __

Dear Grandad,

I like it here. I've got a crab under a rock.

We have been swimm__ __ __ every d__ __.

We went out on Uncle Dan's boat and it was

very windy so we went v__ __y fast.

I got wet with spr__ __. See you on

Sun__ __ __ for lunch.

Love f__ __ __ Clayton

Mr R Green

16 West Road

Haytown

NW4 2RT

Draw Uncle Dan's boat.

★ Clayton got a crab. | Yes | No |

★ Why did Uncle Dan's boat go fast?

____

★ Who got wet with spray?

____

 e

What are the clues which tell you that Clayton is on a holiday?

____

Title: _______________________

Dear Jenny,

14 Parkway, Greentown
27 December

Thank you for my gift. What a surp__ __ __ __!

I have called her H__ __ry, because she has long silky ha__ __. She sits with m__ in my big ch__ __ __ and she plays with a p__ __r of your old socks. She is getting bigg__ __ every day.

Love f__ __ __

Aunt Cl__ __ __

air

st__ __ __

ch__ __ __

H__ __ __y

Cl__ __ __

p__ __ __

surprise box

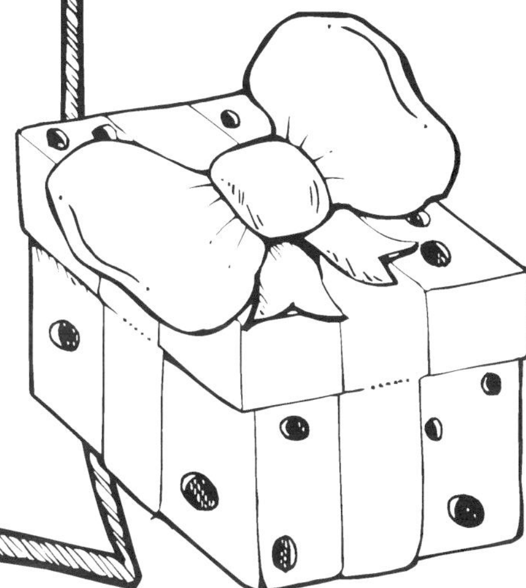

Draw what was in the surprise box.

★ Why did Jenny give her Aunt a gift?

(Clue: Look at the date on the letter.)

____

____

★ Was the gift a baby or fully grown?

Clue: _______________________

(e) Look at the box. What are the holes for? _______________________

____

first	1st
second	2nd
third	___
fourth	___
fifth	___

Minibeasts

Title: ___________________________

What am I?

First clue: I am an insect.

Sec__ __ __ cl__ __: I have long, thin legs.

Th__ __ __ __ __ __ __: I have wings.

F__ __ __th __ __ __ __: My correct name is crane fly.

Fif__ __ __ __ __ __: My name has three parts.
The first part is 'daddy.'

I am a d__ __ __ __-l__ __ __-l__ __ __

Draw me.

★ I have long legs. | Yes | No |

★ I have two names. | Yes | No |

★ What is my correct name?

____

Why do you think I'm called a crane fly? ___________________________

____

Title: ___________________________________

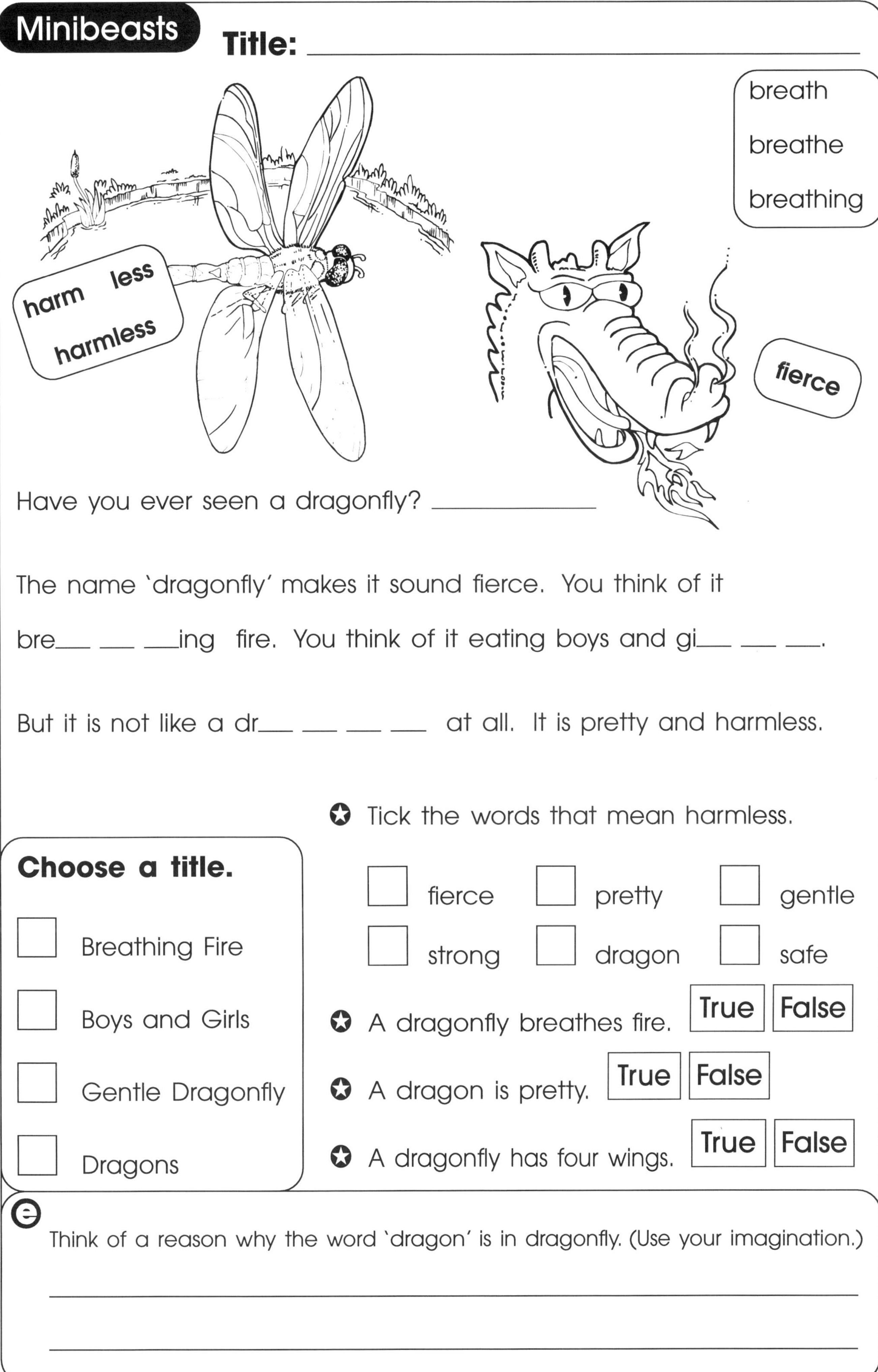

Have you ever seen a dragonfly? _______________

The name 'dragonfly' makes it sound fierce. You think of it

bre__ __ __ing fire. You think of it eating boys and gi__ __ __.

But it is not like a dr__ __ __ __ at all. It is pretty and harmless.

Choose a title.

☐ Breathing Fire

☐ Boys and Girls

☐ Gentle Dragonfly

☐ Dragons

✪ Tick the words that mean harmless.

☐ fierce   ☐ pretty   ☐ gentle

☐ strong   ☐ dragon   ☐ safe

✪ A dragonfly breathes fire. | True | False |

✪ A dragon is pretty. | True | False |

✪ A dragonfly has four wings. | True | False |

℮ Think of a reason why the word 'dragon' is in dragonfly. (Use your imagination.)

____

____

Title: ________________________

first

second

third

prawn

enough

shrimp

Prawns and shrimps are both good to eat.

First you boil them. Second, let them cool and th__ __d, peel the

shells off.

It takes ages to p__ __l enough shrimps for a

meal. Five or six king prawns make a good meal.

⭐ After you boil prawns, what do you do?

____

________________________________ then

____

Choose your title.

☐ Eating Shrimps

☐ A Meal of Prawns

☐ Prawns and Shrimps to Eat

⭐ Why is it easier to make a meal of prawns than shrimps?

____

____

____

🄴 Imagine a nice meal of prawns. What else would you like to have on your plate?

____

Title: ___________________________________

worm

w___ ___k

w___ ___d

w___ ___st

worm

castings

fer til i ser

fertiliser

Plants grow well when worms are in the soil. Wo___ ___ ___ eat all the rotted leaves and dead insects in the s___ ___l.

Worm poo is called 'castings.' Worm cas___ ___ ___ ___ ___ make good fert___ ___ ___ ___ ___ ___ so plants grow better.

W___ ___ ___s are like little gardeners work___ ___ ___ for you below the ground.

Choose a title.

- [ ] Green Plants
- [ ] Underground Gardeners
- [ ] Rotted Leaves
- [ ] Dead Insects

★ What do worms eat? ___________________________

____

★ What are 'castings'? ___________________________

____

e Why do plants grow well when worms are in the soil?

____

____

Title: _______________________________

use

used

using

uses

broken claw

Crabs live under rocks at the beach. They have big claws which they use for eating just like we u__ __ our fingers.

Sometimes cr__ __ __ fight and a cl__ __ is broken off. The crab then grows a new claw.

Choose the best title and write it at the top of the sheet.

- [ ] Crab Claws
- [ ] Fighting Crabs
- [ ] Crabs
- [ ] Under Rocks

✪ What do crabs use their claws for?

1. _______________________________

2. _______________________________

✪ What happens when a crab's claw

is broken? _______________________

____

e Why do you see crabs with two different sized claws?

____

Title: _______________________

Australia has lots of native parrots. One tiny parrot,

which is a favou__ __ __ __ pet, is the budgie.

Flocks of bud__ __ __ __ live wild in Australia.

any

many

favourite

Australia sends its budgies all

over the world.

Now bud__ __ __ __ are

bred in many countries and in

ma__ __ colours.

native

Australia

Choose your title.

- [ ] Australia
- [ ] Tiny Parrots
- [ ] Cages
- [ ] Blue and Yellow

✪ **Draw a budgie in a cage. Give it a perch, swing, ladder, plus seed and water dishes.**

ⓔ The budgie's proper names are 'budgerigar' or 'parakeet'.

Why do you think people call them budgies?_______________________

____

Title: ___________________________

tch

ha__ __ __

ma__ __ __

ca__ __ __

scra__ __ __

take
make
like
pike

tadpoles

We are hat__ __ing from our eggs. Our mother is not here to take

care of __ __.

We are not much big__er than a mat__ __ head.

We make a good cat__ __ for small fish. Some birds even like to

catch and __ __t us.

Three of these are good titles, one is not.

☐ Just Hatched

☐ On Our Own

☐ Matches

☐ A Feed for Fish

⭐ Our mother takes care of us.　True　False

⭐ Our mother is a frog.　True　False

⭐ Small fish eat us.　True　False

ⓔ How do you think a bird would catch a tadpole? ___________________

____

Title: ___________________________

all

call

sm__ __ __er

c__ __ __ed

f__ __ __ing

Starfish are rather odd. We call them fish but they don't look like f__ __ __ at all. We ca__ __ them 'star' but they're not in the sky.

Starfish eat sm__ __ __ sea creatures and coral. They have little sucker pads on their f__ __ __ (5) arms to help them move.

Choose the best title.

- ☐ Sucker Pads
- ☐ Starfish
- ☐ Sea Creatures
- ☐ Coral

★ How many arms does a starfish have?

____

★ Why do we call them starfish?

____

____

e Do you think 'starfish' is a good name? ___________________

Think of another name. _______________________________

Title: _________________________

Something is biting me. Look at all these bi__ __ __. Each bite is red and it__ __y. I think it must be a fl__ __. Fleas are good at bit__ __ __.

Don't look so pl__ __sed. My fl__ __ might l__ __p on to you. Fl__ __s are good at ju__ __ing on their powerful back legs.

Choose a title.

☐ Don't Scratch!

☐ Flea Bites

☐ Powerful Legs

☐ Good Leapers

⭐ Fleas are good at two things.

1. _________________________

2. _________________________

⭐ What do fleas use for jumping?

____

____

e Why do fleas bite people (or animals)? _________________________

____

____

Title: _______________________

break

broken

speak

spoken

A bird has caught a lizard. Poor lizard. But look, the clever liz___ ___ ___ has brok___ ___ off its tail and has run aw___ ___. It's safe, even if it only has a stump of a t___ ___ ___.

The liz___ ___ ___ will now grow a new t___ ___ ___ and the bi___d will have on___y a small lunch.

Choose your title.

- [ ] Poor Lizard
- [ ] New Tail
- [ ] Clever Lizard
- [ ] Small Lunch

✪ Who caught the lizard? _______________

✪ How did the lizard escape? _______________

____

✪ What did the bird have for lunch?

____

e

Why did the author write 'poor lizard' then 'clever lizard'?

____

____

Title: ________________

kangaroo

lk

mi__ __

si__ __

ta__ __

wa__ __

pouch

world

I'm a baby kang__ __ __ __.

When I was born, I was only this big ◯ . I suck __ __lk from my

mother and I live in her pouch. You can often see my l__g__ sticking

out of her p__ __ __ __.

I live in Aus__ __ __ __ __ __ but

you can also see me in zoos all

over the w__ __ __ __.

Choose a title.

☐ Long Legs

☐ World Zoos

☐ Mother's Milk

☐ Baby Kangaroo

⭐ Baby kangaroos ride in their mother's pouch. | True | False |

⭐ What do kangar__ __ __ drink?

____

⭐ A baby kangaroo walks and talks. | True | False |

e Why do you think baby kangaroos travel in the pouch? ________________

____

Title: ___________________________________

ou

m__ __nd

s__ __nd

gr__ __nd

r__ __nd

f__ __nd

Termites are soft and white. They live in big families in the dark under the gr__ __nd. They can make large m__ __nds above gr__ __nd. The m__ __nd is full of tun__ __ __ __ and dark rooms to live in.

Termit__ __ eat wood. They eat trees, posts and even the w__ __d in your house.

Choose a title.

☐ Wooden Posts

☐ Termites

☐ Tunnels

☐ Mounds

★ Termites always live underground.
| True | False |

★ Termites live in the dark. | True | False |

★ What do termites eat? ___________________________

e

Termites turn dead trees back into soil. Is this good or bad? ___________________

Why? ____

____

Title: ___________________________

busy

active

speedy

fast

hurry

pro tec ted

protected

wing cases

Beetles always seem to be very busy. They run everywhere at top

sp__ __d. Perhaps they think they will be eat__ __ by a passing bird.

Beet__ __ __ can fly too, but you don't often see their w__ __ __s

because th__ __ are pro__ __ __ __ __ __ by hard wing cases

on their backs.

Choose a title.

☐ Wing Cases

☐ A Passing Bird

☐ Speedy Beetles

☐ Busy Bees

• Think of a title.

✪ Beetles are slow insects. |True| |False|

✪ Beetles are eaten by birds. |True| |False|

✪ Beetles wings are hidden underneath wing cases. |True| |False|

✪ What does 'protected' mean?

____

e

What do you think we mean when we say that someone is 'beetling about'?

____

Title: ________________________

pop u lar

popular

old

g__ __ __

h__ __ __

c__ __ __er

f__ __ __ing

re flect

reflect

I'm a g__ __ __fish. I live in a tan__ of c__ __ __ water.

As I swim you can see g__ __ __ flashes as my scales reflect light.

I'm a pop__ __ __ __ fish to keep. Sometimes I'm kept in a fish

po__ __ in the garden.

Choose 2 titles.

☐ Fish Tank

☐ Flashes of Gold

☐ In the Garden

☐ Goldfish

✪ Write two places to keep goldfish.

________________ and ________________

✪ What does light reflect off? ________________

____

✪ A goldfish needs hot water. | True | False |

e

Why do you think goldfish are popular? ________________

____

Title: _______________________________

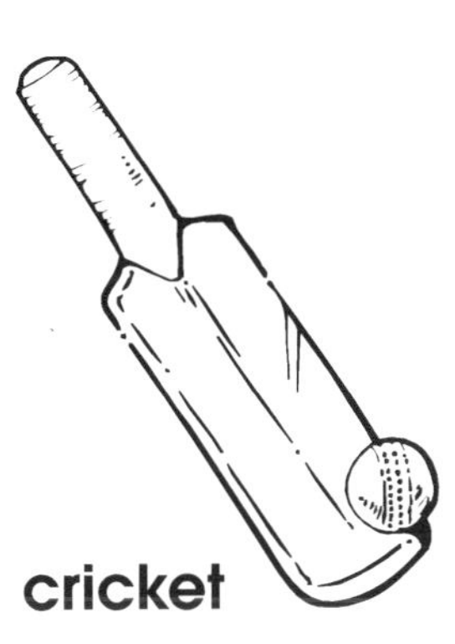

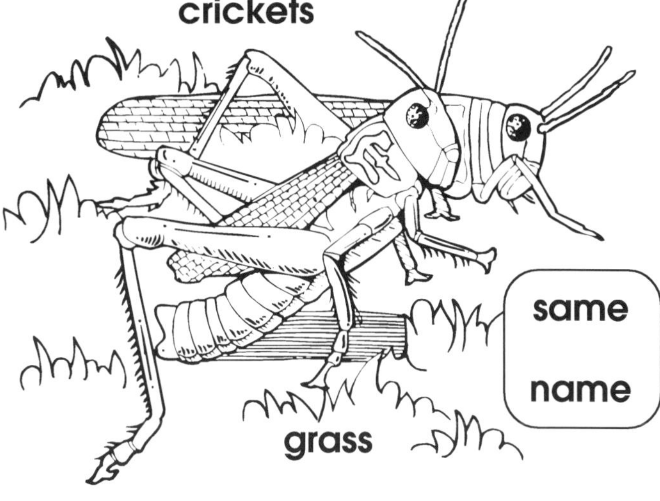

ass
gr__ __ __
p__ __ __
cl__ __ __
br__ __ __

Cricket is played on gr__ __ __. Crickets live on gr__ __ __ (and other plants). The name's the same!

Cri__ __ __ __s are insects so they have six legs, and wings for fly__ __ __. They are in the same cl__ __ __ as gr__ __ __hoppers.

They make their crrr, crrr, crrricket noise by rubbing their legs on th__ __r hard wing cases.

Draw what the story is about.

✪ The story is about cricket bats.

| True | False |

✪ Crickets are in the same family

as _______________________________

✪ Crickets have wings.

| Yes | No | Doesn't say |

Why do you think crickets make that noise? _______________________

____

Title: ______________________

hive

___ve

live

gi__ __

ha__ __

sa__ __

hi__ __

col lect ing
collecting

sugar water

Bees can liv__ in a hole in a tree. They can also l__ __ __ in a beeh__ __ __ made of wood.

Bees go out col__ __ __ __ing honey. Some of the hon__ __ is food. Some is sav__ __ for winter.

If we take the hon__ __, we have to gi__ __ the b__ __s some sugar for winter.

Choose the best title.

☐ A Wooden Hive

☐ Honey Bees

☐ Honey for Winter

☐ Hole in a Tree

⭐ A hole in a tree can be lived in by bees. | True | False

⭐ What do bees use honey for?

____

⭐ Why do we give them sugar?

____

ℯ Why do you think bees save honey for winter? _______________________

____

Title: ___________________________

underground

underneath

undersea

understand

find

mind

bl__ __ __

beh__ __ __

Moles dig long underg__ __ __ __ __ tunnels with their strong claws. It's dark in the tun__ __ __ so they fi__ __ their way by smelling, not by looking. Moles are bli__ __.

Moles eat small bugs and worms. They make rooms undern__ __ __ __ the ground for their babies which drink their mother's milk.

Choose your title.

☐ Mother's Milk

☐ Bugs and Worms

☐ Blind Moles

☐ Underground

✪ What do moles use to dig their tunnels?

____

____

✪ What do baby moles feed on?

____

✪ Can moles see? | Yes | No | Doesn't say |

e Why is 'understand' the odd word in the top left-hand box?

____

____

Title: _______________________

| gar den er |
| gardener |

nch

lu__ __ __

bu__ __ __

mu__ __ __

cru__ __ __

Fre__ __ __

Poor me. I seem to be eaten or stamped on.

Ducks love to e__ __ me.

French people l__ __ __ to eat __ __.

Some birds have me for lu__ __ __.

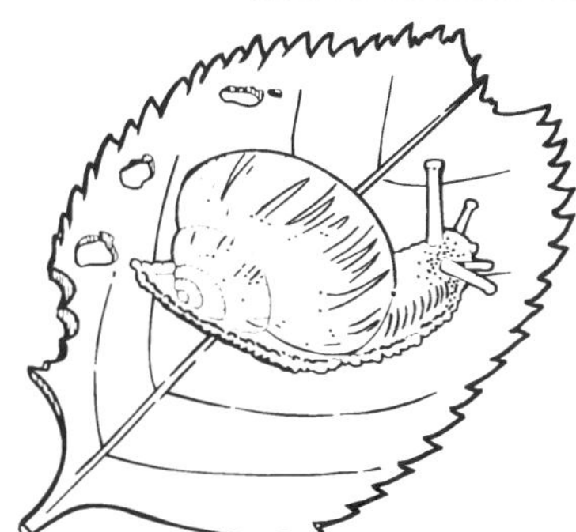

I lov__ to mu__ __ __ on plants, so

gar__ __ __ __ __s poi__ __ __ me.

They also stamp on me with a

cru__ __ __ of my shell. Poor me.

★ Why do ducks love me?

____

____

★ Why do gardeners hate me?

____

____

Think up two good titles.

____

____

____

____

e Do you think it's a good thing to poison snails? _______________________

Why? _______________________

____

Title: ________________________________

Scorpions are like spiders because they have eight (8) legs.

They are also like spiders because th__ __ both eat in__ __ __ __s.

Another way scor__ __ __ __ __ and sp__ __ __ __ __ are the same is that th__ __ both poison their pr__ __.

But sco__ __ __ __ __ __ have a pois__ __ __ __s sting in their tails, while sp__ __ __ __ __ have a poisonous bite.

✪ Write three ways scorpions are like spiders.

____

____

____

✪ How are scorpions different to spiders?

____

____

____

Think of a good title.

____

____

____

____

Make it less than six words.

e Can you think of other ways that spiders and scorpions are different?

____

____

Title: _______________________________

I have a shell but I'm not an egg.

I can be a pet b__ __ I'm no__ furry.

I will eat plants in your ga__ __ __ __ but __'__ n __ __ a snail.

I can live to be very old b__ __   I'__ __ __ __ your grandfather.

My cousin is a turtle and my name starts with a T.

I am a t__ __ __oise.

Choose the best title.

- [ ] Eggshells
- [ ] Turtle
- [ ] Grandfather
- [ ] Shelly

★ Write things which I'm not.

I'm not _______________________________

I'm n__ __ _______________________________

I'__ __ __ __ _______________________________

__'__ __ __ __ _______________________________

e How is a turtle different from a tortoise? _______________________________

____

Title: ________________________________

enough

ever

n___ ___ ___ ___

for___ ___ ___ ___

___ ___ ___ ___y

answers

Question: Do you know why a centi___ ___ ___ ___ never wears sh___ ___s?

Here are some answ___ ___s. Think of some of your own.

Answer: The sh___ ___ shop never had en___ ___gh pairs in the same size.

Answer: It would take forev___ ___ to get dressed in the morn___ ___ ___.

Answer: It liked to wear football boots instead.

Your answer: ________________________________

____

Choose a title.

☐ A Hundred Shoes

☐ Centipede Riddles

☐ Football Boots

☐ Shoe Shop

shoes

e Think of a riddle about centipede legs.

Question: ________________________________

Answer: ________________________________

Minibeasts

Title: _______________________________

_o_e

p__k__

str__k__

m__r_

n__s__s

there   over there

their   their fur

they're   they are

Guin__ __ p__ __s make good pets.

Th__ __'re warm and furry and like to be str__ __ __d.

Th__ __ __ fur can be white, black or brown. Th__ __ __ little

black nos__ __   p__k__ here and th__ __ __ looking for food.

Th__ __'re easy to feed.

Give them th__ __ __ food of seeds,

greens or bread and it'll be eaten and

th__ __'ll be looking

for m__r__.

✪ What do guinea pigs eat? _______________

____

✪ What does a guinea pig feel like?

____

Think of a good title.

____

____

e If you lived in a flat, would a guinea pig be a good pet? _______________

Why? _______________________________

____

Title: _______________________________

oy

R__ __
t__ __
b__ __
j__ __
enj__ __ed
R__ __al

standard   toy

When R__ __ was nine he was given a puppy for his birthday.

R__ __'s pup was not like other dogs. She had wool, not hair.

She was a p__ __ __le. Some p__ __ __ __ __ __s are large.

They are called st__ __ __ard poodles. Some are tiny. They are

called t__y poodles. All p__ __ __les enj__ __ running and

playing with balls. R__ __'s poodle was a t__ __.

Choose a title.

☐ Roy's Puppy

☐ Poodles

☐ Woolly Coat

☐ Toy

★ Was Roy's puppy a boy or a girl? _______________

What word was your clue? _______________

____

★ Would Roy's puppy grow very big? _______________

What was your clue? _______________

____

ⓔ

Which size poodle would you like? _______________________

Why? _______________________________________

____

Title: ___________________________

Hermit crabs are a bit like spiders because they have eight (8) legs.

They are a bit like snails bec__ __ __ __ they carry their houses

on th__ __ __ backs.

Snails can grow bigger shells as they grow bigg__ __ but hermit

crabs can't. They have to go looking for a big__ __ __ shell that

is empty.

✪ How many legs has a hermit crab?

____

✪ How are snails and hermit crabs the

same? ___________________________

____

✪ Why do you think a shell would be

empty? __________________________

____

Choose the best title.

☐ Crabs and Spiders

☐ Snail Shells

☐ Hermit Crabs

☐ Crab Legs

ℯ When you see hermit crabs in a group, they all look different. Why is that?

____

Title: _______________________

This is a poem for you to finish.

I have strong back legs to j__ __ __ with.

I can hop a long, l__ __ __ way.

I have short fr__ __ __ legs to cl__ __ __ with.

Round my pond I like to st__ __.

I must br__ __the in air like you do.

On my head I have big ey__ __.

I lay lots of e__ __s like fish do.

I like eating bugs and fl__ __ __.

Think up two titles.

____

____

____

✪ A frog uses its back legs for

____

and its front legs for _______________________.

✪ What does a frog like to eat?

____

ℯ Why do you think a frog has eyes on the top of its head?

____

Title: _______________________

they

grey

ob__ __

pre fer
prefer

qu

quite

quail

__ __eer

Quails are quite small birds.

They eat all sorts of things

like seeds, small bugs and grass.

One que__ __ thing about quails is that they don't often fly.

They can fly but they pre__ __ __ being on the ground.

Because th__ __ are mostly brown and gr__ __, they are hard to

see if th__ __ stay still.

★ Your turn! Quails can be:

☐ green ☐ brown

☐ red ☐ purple

☐ grey ☐ yellow

Where do you think a quail lays its eggs?

**Here are some long titles.
Tick the best one.**

☐ Seeds, Small Bugs
and Grass

☐ Stay Still and
We Won't See You

☐ Queer Birds
Who Don't Like to Fly

🅴 Why do you think quails don't fly much? _______________________

____

Title: _______________________________

noc tur nal
nocturnal

insects

fruit

igh

l__ __ __t
n__ __ __t
dayl__ __ __t
fr__ __ __tened

some
come

What are we? Here are your clues.

- We are warm-blooded and furry and we can fly.

- We are noc__ __ __ __ __ __ which means we come out at n__ __ __t and are asleep in dayl__ __ __t.

- Some people are fr__ __ __tened of us.

- Some of us eat insec__ __ and some eat fr__ __ __.

- We hang upside down to sl__ __p.

  We are __ __ __ __.

Draw us sleeping.

★ We are insects which eat fruit. | True | False |

★ We have fur on our bodies. | True | False |

★ What does nocturnal mean?

____

e

How many clues did you need?________ Which clue gave you the answer?

____

____

Title: _______________________

wh

w__en

__ __y

__ __at

__ __ing

__ __ee

__ __ere

__ __ich

Whing, whee, w__ing, __ __ee! I'm in your bedr__ __m.

But __ __ere? I'm fly__ __ __ around your head. Slap!

You missed me. W__ere am I now?

You can't feel me l__ __d on you. Now I'll pi__ __ __ __ your

skin and suck a bit of your blood. Scratch, scr__ __ __ __,

off I fly until you fall asleep again. What a night!

One title is poor. Which one?

☐ What a Night!

☐ Whing, Whee

☐ Bedroom

☐ Scratch, scratch

✪ Have you had a mosqu__ __ __ in your room? _______________________

✪ What noise did it make?

____

✪ What did it do when you were asleep?

____

e

Why do you think a mosquito needs your blood? _______________________

____

____

Title: _______________________

butter

A butterfly is a pretty insect. But what a funny name!

A fly made of butter?

Butter is yellow and some but__ __ __f__ __ __s are

yel__ __ __, but some are blue or white or red.

But__ __ __ is soft and spr__ __d on br__ __d. It would be

funny to sp__ __ __ __ a butterf__ __ on bread.

An animal who butts with its h__ __d is a goat. It is a butter.

A butterfly never butts.

Choose a title.

☐ Yellow Butter

☐ Bread Spread

☐ Butting Goat

☐ A Funny Name

✪ What colours are butterflies?

____

✪ A goat is spread on bread. | True | False |

e

How do you think the butterfly got its name? _______________

____

Title: ___________________________

eight	8
___________	9
twelve	___
___________	16
eighteen	___

This is my head.

breathing

What am I?

What am I?

- I have twel__ __ eyes.

- I h__ __ __ __ __ __ __ __ (8) legs on each side of my body.

- I h__ __ __ __ __ __ __ (9) breathing holes on each s__d__.

- I hatch out of an egg and I eat leaves. As I grow fatter,

  I shed my skin and gr__ __ a big__ __ __ one.

- I change into a moth or a butterfly.

★ I am a __ __ __ __ __ __illar.

Choose a title.

☆ How many legs?   ______

☐ Moths

☆ How many eyes?   ______

☐ Breathing Holes

☆ How m__ __y breathing holes? ______

☐ Caterpillar

☆ What will I be when I'm grown up?

☐ Butterfly

____

Ꙫ If you were a caterpillar, what would you be scared of?

____

Minibeasts

Title: _______________________

ould

c_ _ _ _ _

w_ _ _ _ _

sh_ _ _ _ _

dif fer ence
difference

guess
guessed
gnaw
gnawing
know
kn_ _ing

What am I?

What am I?

- I have four legs. (It's too soon to guess yet.) **Yes** **No**

- I have a furry body and whiskers.
  (Should you gu_ _ _ yet?) **Yes** **No**

- I have sharp front teeth for gnawing which keep
  growing all my life. (Are you sure yet?) **Yes** **No**

- I eat seeds and anything else I can find.
  (Would you make a gu_ _ _ yet?) **Yes** **No**

- I often live in houses and traps are set for me.
  (Now are you sure?) **Yes** **No**

- I am a _ _ _ _ _ .

✪ I have no fur. **True** **False**

✪ What's the difference between
my teeth and yours?

____

____

**Draw me when you have
finished the story.**

e Did you guess right away or did you read all the clues first? _______________

Why? _______________________

Minibeasts

Title: ___________________________

other

m__ __ __ __ __ __

br__ __ __ __ __ __

an__ __ __ __ __ __

What am I?

- I am small and pink. (Too soon to guess?)   **Yes** **No**

- I have warm blood and very little hair.
  (Is it sensible to guess yet?)   **Yes** **No**

- I'm a baby animal with lots of brothers
  and sisters. (Enough information yet?)   **Yes** **No**

- I have a curly tail and drink my mother's m__ __k.
  (Do you know yet?)   **Yes** **No**

- When I grow up I might be bacon or pork.
  (Now you know!)   **Yes** **No**

I am a __ __ __ __ __ __!

Draw me when you have finished the story.

★ I am grown up.   **True** **False**

★ I am warm-blooded.   **True** **False**

★ I come from a big family.   **True** **False**

How many clues did you need to be sure you were correct? ___________

Which clue gave it away? ___________

Title: ___________________________

would

c__ __ __ __

sh__ __ __ __ __

What are we?

What are we?

guess

maybe

perhaps

often

sometimes

- We are insects, so we have

  s__ __ legs. (Should you guess yet?)   **Yes** **No**

- We are often found in houses.

  (Wou__ __ you guess now?)   **Yes** **No**

- People don't like us. They say we're dirty.

  (Does that help you?)   **Yes** **No**

- Peo__ __ __ spray us, hit us and stand on us.

  (Perhaps you can gu__ __s now!)   **Yes** **No**

- Our name has two parts.

  (a) A male hen. __ __ __ __

  (b) Like coach but starts with 'r' __ __ __ __ __

  We are c__ __ __r__ __ __ __es.

SPRAY

Draw us when you have finished the story.

✪ How many legs do we have?

____

✪ Why don't people like us?

____

✪ How do they try to get rid of us?

____

ℯ Think of some reasons why cockroaches are still with us if we keep attacking them.

____

____

Title: _______________________

sens ible
sensible

What am I?

people

tch

ha______ ______ ______
ca______ ______ ______
fe______ ______ ______
hu______ ______ ______

What am I?

- I have just ha___ ___ ___ed
  from an egg. (Should you guess yet?)   Yes No

- I'll run with my brother and sisters down to
  the river. (Is it sensi___ ___ ___ to guess yet?)   Yes No

- Fish and birds c___ ___ ___ ___ us.
  Lots of us are eaten. (Does that help you?)   Yes No

- Some people make our skins into belts and
  handbags when we grow up. (Guess now?)   Yes No

- I am in the same family as the crocodile.

- I am a baby al___ ___ ___ ___ ___ ___r!

Draw me when you have finished the story.

✪ I am an adult.   True False

✪ We are food for some fish and birds.
   True False

✪ My skin makes good belts.
   True False

ℯ

Are the baby alligators looked after by their mother? _______________________

What were the clues that helped your answer? _______________________

____